CREEPY BUT COOL

Alan Walker

A Crabtree Seedlings Book

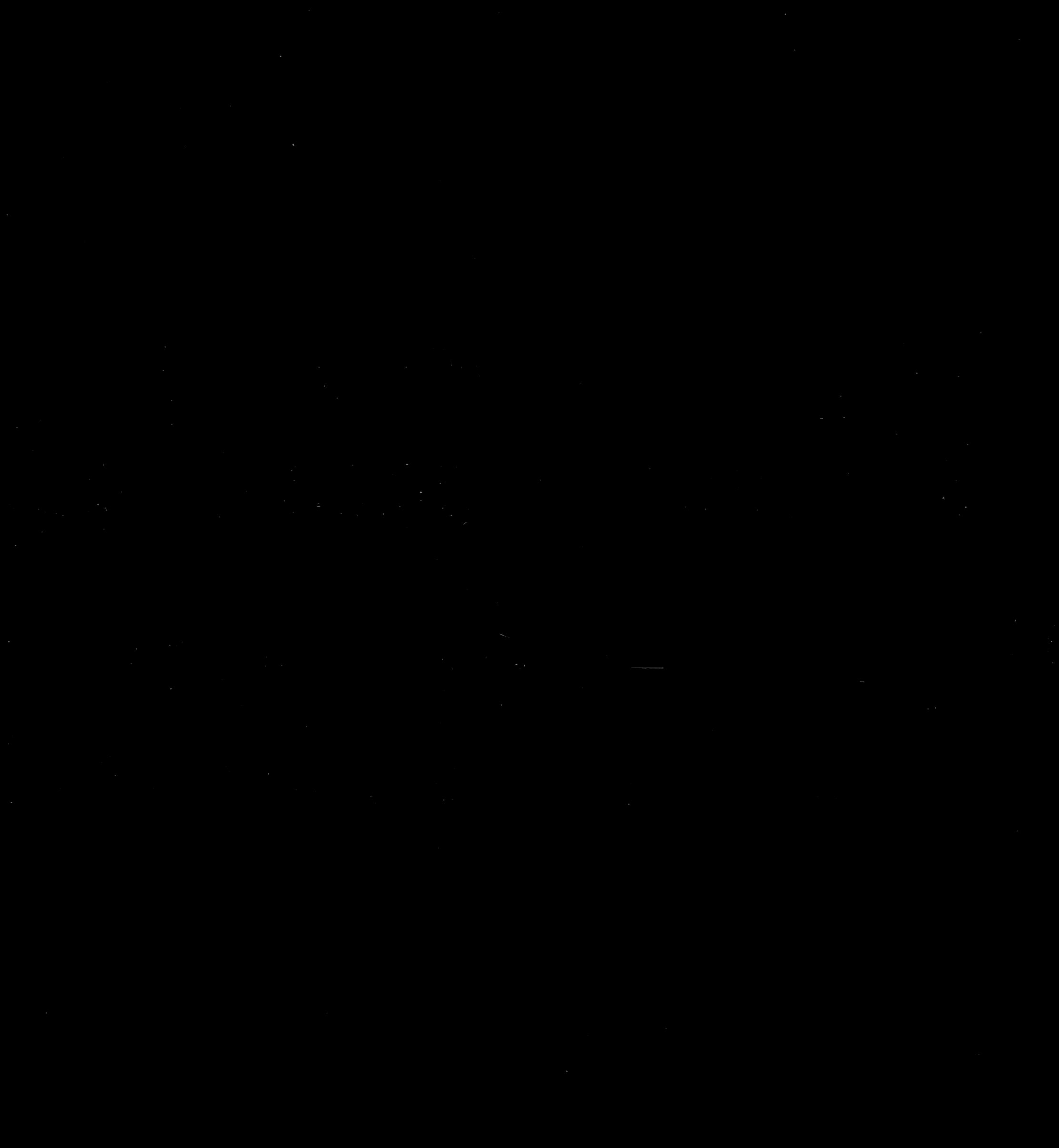

TABLE OF CONTENTS

INSECT, BUG, OR BOTH?

Scientists put living things into six groups called *kingdoms*.

The Six Kingdoms:

Archaebacteria

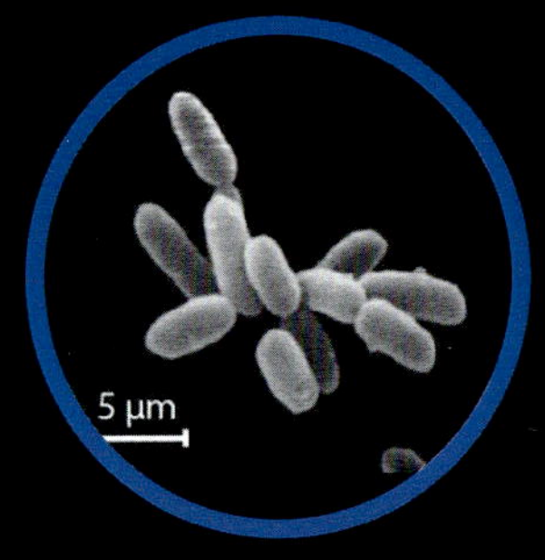

Eubacteria

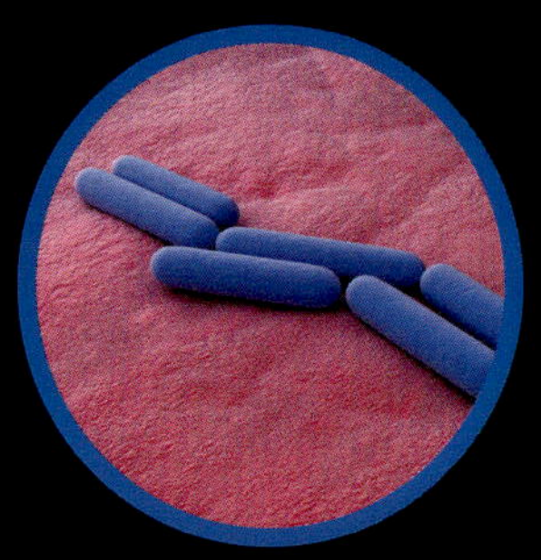

Protista

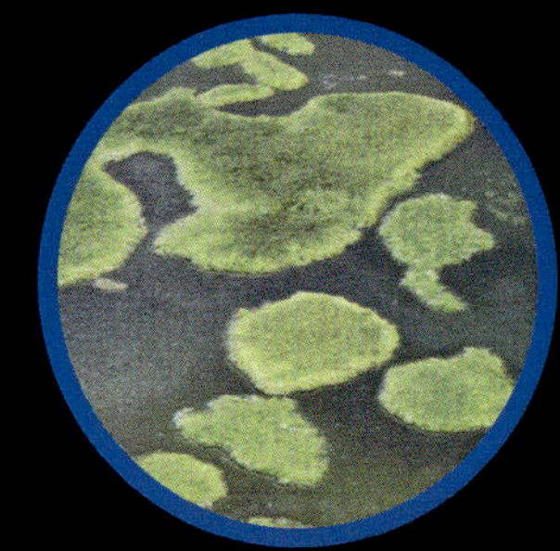

Fungi

Plants

Animals

Each kingdom is separated into smaller and smaller groups.

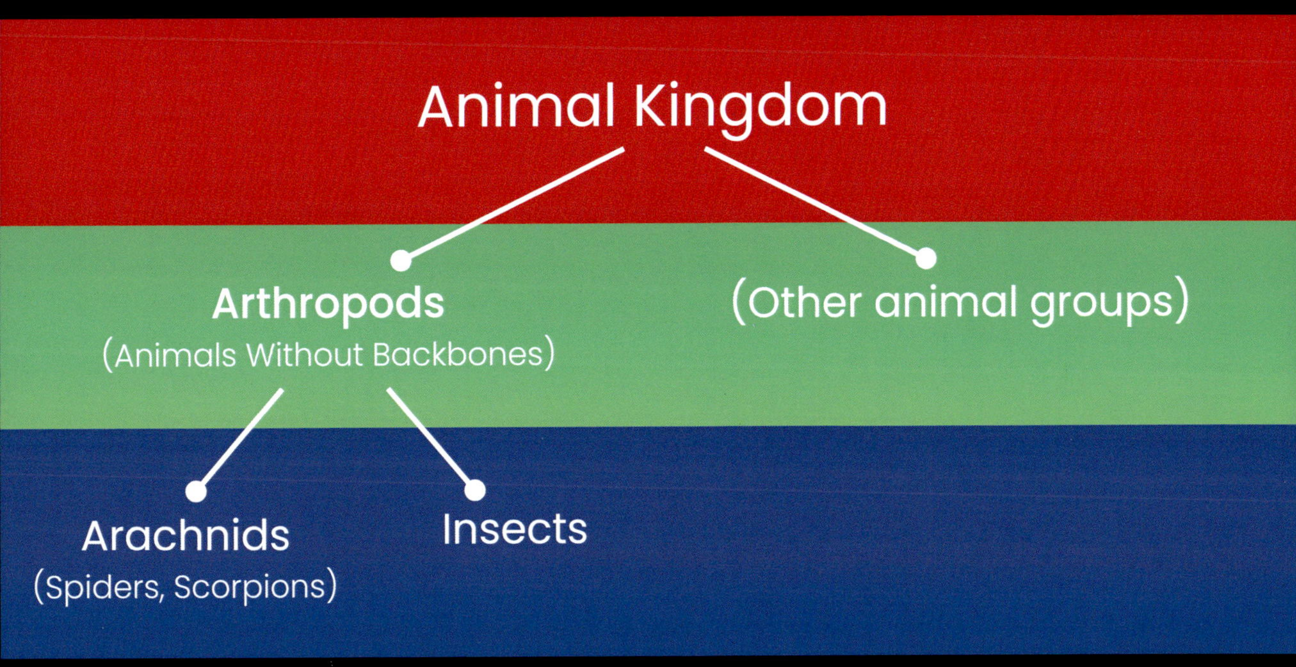

This chart shows just a few of the groups in the Animal Kingdom.

Insects are grouped together because they have six legs, a hard outer covering, and two antennae.

CREEPY OR COOL?

About 300 million years ago, there were dragonfly-like insects that had wingspans measuring almost 3 feet (1 meter)!

Meganeura **was a kind of insect that looked a lot like a giant dragonfly. Some were the size of a large hawk!**

Bug is a word people often use for insect
But scientists say only some insects can be called bugs. The difference is in the mouthparts . . .

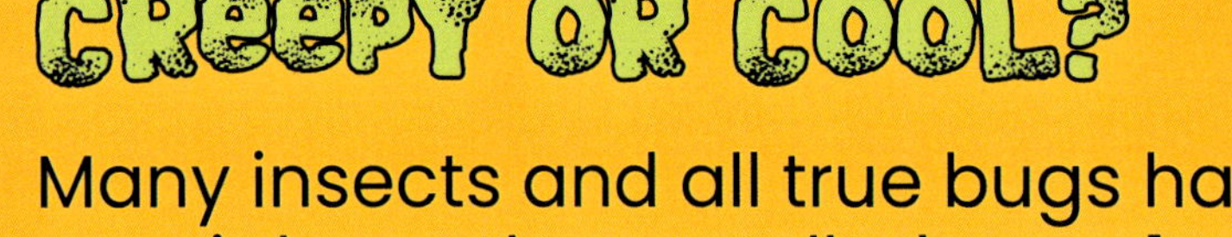

Many insects and all true bugs have a special mouthpart called a **proboscis**. True bugs use this to stab food and suck up the juices. True bugs cannot roll up their proboscises—other insects can.

A butterfly can roll up its proboscis.

A true bug cannot
ll up its proboscis.

Assassin Bugs

There are many kinds of assassin bugs. All are **predators** that feed on other insects, spiders, millipedes—and sometimes humans!

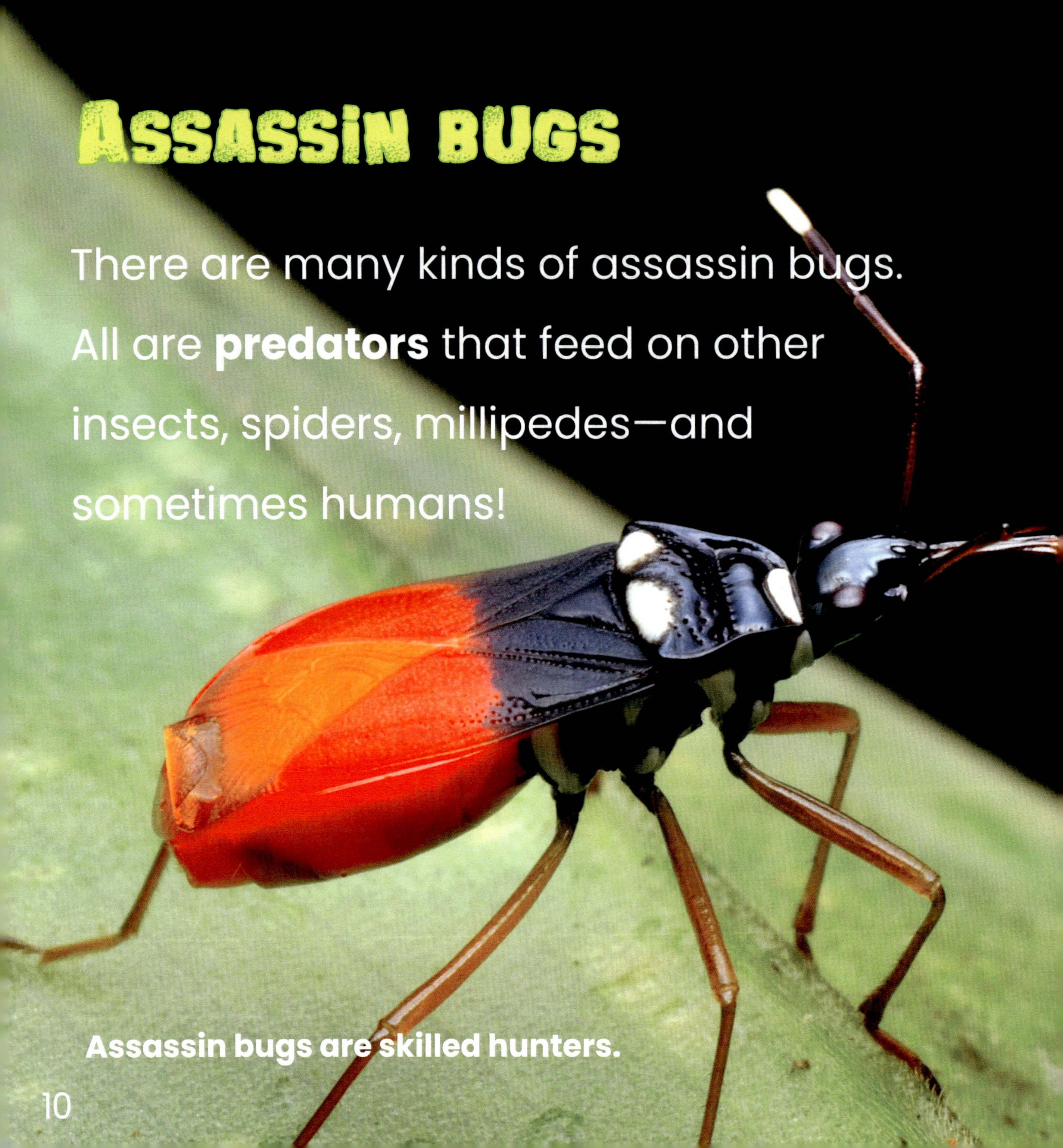

Assassin bugs are skilled hunters.

CREEPY OR COOL?

The masked hunter bug covers itself in dust or lint. This **camouflage** helps it hide from other predators or **prey**.

Wheel bug saliva **paralyzes** prey and turns its prey's insides into a runny soup! Soupy insides are easier to suck up!

A wheel bug uses its proboscis to inject paralyzing saliva.

Kissing bugs bite animals and people and suck their blood!

CREEPY OR COOL?

We call them kissing bugs because they bite their victims' faces around the mouth and eyes.

BUGS THAT SMELL REALLY BAD!

brown marmorated stink bug

When threatened, stink bugs give off a stinky smell to defend themselves.

CREEPY OR COOL?

Stink bugs have special body parts on the underside of their thorax. These body parts produce the bad smells that scare away predators.

As well as using bad smells, some stink bugs have other ways to defend themselves.

A green stink bug uses camouflage to hide on a leaf!

The harlequin bug's bright colors warn predators to stay away!

CREEPY OR COOL?

The peanut-head bug has large, false eyes on its hind wings to scare away predators.

Skaters, Swimmers, and Giants

Many true bugs live in ponds, rivers, and streams. These bugs move around in different ways.

Water striders, also known as pond skaters, have hairs on their feet that allow them to stand on, or skate across the water's surface.

CREEPY OR COOL?

Unlike real scorpions, a water scorpion uses its tail like a snorkel, not a weapon.

Water scorpions use their rear legs like oars to move through the water.

Giant water bugs can grow up to 6 inches (15 centimeters) long. They feed on insects, fish, frogs, and **crustaceans**.

In some countries, people snack on giant water bugs.

Glossary

camouflage (KAM-uh-flazh): Colors or patterns that help animals blend with their surroundings

crustaceans (kruh-STAY-shuhnz): Sea creatures that have exoskeletons, such as crabs, lobsters, and shrimp

paralyzes (PA-ruh-lize-iz): Makes something, or someone, unable to move

predators (PRED-uh-turz): Animals that hunt and eat other animals

prey (PRAY): Animals that are hunted and eaten by other animals

proboscis (pruh-BAHS-iss): The tube-like mouthpart that insects use for feeding

School-to-Home Support for Caregivers and Teachers

This book helps children grow by letting them practice reading. Here are a few guiding questions to help the reader build his or her comprehension skills. Possible answers appear here in red.

Before Reading

- **What do I think this book is about?** *I think this book is about creepy bugs. I think this book is about bugs' defenses.*
- **What do I want to learn about this topic?** *I want to learn more about why some bugs sting people. I want to learn the different categories of bugs.*

During Reading

- **I wonder why...** *I wonder why some people call insects bugs. I wonder why some bugs feed on other bugs.*
- **What have I learned so far?** *I have learned that some bugs cover themselves in dust or lint to help them hide from predators. I have learned that the saliva of some bugs paralyzes their prey.*

After Reading

- **What details did I learn about this topic?** *I have learned some bugs bite animals and people and suck their blood. I have learned that stink bugs give off a stinky smell to defend themselves from predators.*
- **Read the book again and look for the glossary words.** *I see the word **proboscis** on page 8, and the word **camouflage** on page 11. The other glossary words are found on page 23.*

Library and Archives Canada Cataloguing in Publication

CIP available at Library and Archives Canada

Library of Congress Cataloging-in-Publication Data

CIP available at Library of Congress

Crabtree Publishing Company

www.crabtreebooks.com 1–800–387–7650

Print book version produced jointly with Blue Door Education in 2022

Written by Alan Walker

Print coordinator: Katherine Berti

Printed in the U.S.A./062021/CG20210401

Photo Credits: Cover photo ©shutterstock.com/Jirasak Chuangsen, Page 5 large beetle ©shutterstock.com/Krissanakorn Phadungkarn, other images Sebastian Kaulitzki, dominique landau, Nicky Rhodes, Light & Magic Photography, worldswildlifewonders, Page 6 ©shutterstock.com/Kirsanov Valeriy Vladimirovich, Page 7 meganeura ©shutterstock.com/ Warpaint, dragonfly silhouette ©shutterstock.com Algonga, human silhouette © Michal Sanca, Page 9 wheel bug ©shutterstock.com/Gerry Bishop, butterfly ©shutterstock.com/Olga Bogatyrenko, Pages 10-11 ©shutterstock.com/RAMLAN BIN ABDUL JALIL, Page 11 inset photo ©shutterstock.com/D. Kucharski K. Kucharska, Page 12-13 ©shutterstock.com/Michael G McKinne, Pages 14-15 ©shutterstock.com/schlyx, Pages 16-17 ©shutterstock.com/Marco Uliana, Page 18 ©shutterstock.com/Alonso Aguilar, Page 19 ©shutterstock.com/Peter Yeeles, inset photo ©shutterstock.com/COULANGES, Page 20 ©shutterstock.com/mjf99, Page 21 ©shutterstock.com/ Kirsanov Valeriy Vladimirovich, Page 22 © AndrewASkolnick, page 22 ©shutterstock.com/Pheobus, inset photo ©shutterstock.com/nicemyphoto; archaebacteria page 4, courtesy of NASA

Published in the United States
Crabtree Publishing
347 Fifth Ave.
Suite 1402-145
New York, NY 10016

Published in Canada
Crabtree Publishing
616 Welland Ave.
St. Catharines, Ontario
L2M 5V6